YOUR KNOWLEDGE HAS VALUE

- We will publish your bachelor's and
  master's thesis, essays and papers

- Your own eBook and book -
  sold worldwide in all relevant shops

- Earn money with each sale

Upload your text at www.GRIN.com
and publish for free

Nilofar Khan Roshina Bashir

Nutritional Deficiencies of Adolescent Boys and Girls of Kashmir Valley (J&K, India)

GRIN Publishing

Bibliographic information published by the German National Library:

The German National Library lists this publication in the National Bibliography;
detailed bibliographic data are available on the Internet at http://dnb.dnb.de .

Imprint:

Copyright © 2014 GRIN Verlag GmbH
Print and binding: Books on Demand GmbH, Norderstedt Germany
ISBN: 978-3-656-71752-2

NUTRITIONAL DEFICIENCIES OF ADOLESCENT BOYS AND GIRLS OF KASHMIR VALLEY (J&K, INDIA)

Nilofer Khan

Sr. Professor, Institute of Home Science, Faculty of Applied Science & Technology, University of Kashmir, Srinagar, India

ABSTRACT

The present study focused on nutrition deficiencies of adolescent boys and girls. The field sample for the present study was undertaken in six districts of Kashmir valley J&K (viz, Srinagar, Budgam, Anantnag, Kupwara, Pulwama and Baramulla), covering a sample of 1500 adolescents i.e. 750 boys and equal number of girls in the age group of 10-19 years, study in Government Schools. The findings of the study suggests that lifestyle including nutritional habits track from adolescent into adulthood, thereby lead to increased incidence of chronic disease e.g. cardiovascular diseases, diabetes and cancer. Exposure in childhood and adolescence to adverse life style and faulty food habits such as poor food intake, special meal preferences and patterns and above all sedentary life style may exacerbate this, thus worsening the prognosis. Hence promotion of healthy nutrition habits and physically active life style during adolescent period is a critical public health strategy.

Keywords: Nutrition, Deficiency, Adolescents, Boys, Girls

INTRODUCTION

Adolescence is a time of change to adult behavior and there by eating habits of childhood gradually change into those typical of an adult. Adolescence is, therefore, an important time that demands for health and nutrition education. Eating habits may be erratic large quantities may be eaten one day and very little next day. It has been pointed out by researchers that adolescents in different parts of the country had nutritional deficiencies. It occurs in boys as well as in girls. Adolescent girls are at special nutritional risk because of iron deficiency anemia. The requirement of iron, which is 18 mg/day, is needed not only to make losses due to menses but also to build up reserves. Calorie requirement for most adolescent is high. An adolescent may rush off to school without eating breakfast. When away from home he or she usually eats readily available meals that are acceptable to its peer groups. This means snacks in the form of fast-food (junk food). He/she eats fewer meals at home where parents can provide them nutritious diet. Adolescents may indulge in food fads, macrobiotic diets and semi starvation regimens in calories, vitamins and minerals. An adolescent protein need/unit body weight is higher than that of adult but less than a rapidly growing infant. 7 Introduction Adolescents have higher vitamin and mineral needs compared with people at most other life stages. Adolescents are mostly concern with vitamin A, calcium and iron each of which plays an important role in growth. Adolescents who do not achieve sufficient bone density have greater risk of developing osteoporosis later in life (Heald 1975; Thomas 1989). Physical changes cause an adolescent to focus attention on his body as he tries to incorporate his new appearance into his developing sense of identity. Many adolescents go through stages, which they are pre-occupied with their appearance and body functions. They may see nutrition as helpful or harmful to their developing body image. A boy may show concern about his body in relation to athletic ability. He may want to eat more to increase his weight and muscle mass. The deposition of fat which normally occurs in adolescent girls may cause her to become concerned that she is getting fat limiting calorie intake at this time may interfere with linear growth. The onset of obesity during adolescence may contribute to a number of psychological problems. It may interfere with development of positive body image (Starz 1983; Kapil 2002).

Dietary recommendations during adolescence must take into account the social and attitudinal characteristics of the individual as well as the timing and the rate of growth. Greater independence from family supervision and guidance is associated with increased peer conformity and influences of mass media. Rapid changes in body create alterations in body image and individual reactions to those changes. Emotional instability may cause intermittent stress. Physical activity may be higher among individuals who participate in competitive sports but very low in those with sedentary pursuit's time schedule may lead to the omissions of some meals or to greater frequency of eating may be consumed more often away from home and may commonly be bought in franchised food outlets. Interest of non-traditional eating pattern may increase. Nutrient needs during adolescence are dictated by the rate of growth. Requirement increases at the outset of growth. Spirit reaches their maximum at the time of peak growth and gradually approach adult levels as growth subsides (Srilakshmi 2002).

The age of puberty has shifted gradually to earlier years and a lengthened period of education and dependence has served to expand the life of adolescence (Garg 2002). During this period of life the stress of rapid growth becomes evident and are manifest both in Physical and mental changes and in sexual maturation of the individual. If there is non acceptance of adolescent role as an individual or a male or a female, there is opt to develop a defiant loneliness, feeling of rejection, and an unbearable resentment or even hatred of oneself or of one or both parents that may transfer to all members of a particular sex or social group or to society as a whole with disastrous consequences (Bhattacharya 1985). Out of this turmoil there may develop panoply of special medical and social problems. Some are related to genetic factors and some to physiological changes within the individual where as others develop from the search for identity and purpose with concomitant rebellion against a real or apparent restraining society not infrequently the causes interrelated and mutual reinforcing such as defiance of parents, drinking alcohol, drugs, pregnancy and venereal diseases.

MATERIAL AND METHODS

The present study was carried out in Government Schools of six Districts of Kashmir Valley were Srinagar, Budgam, Anantnag, Kupwara, Pulwama and Baramulla. The study was undertaken on 1500 adolescents aged 10-19 years of age, both boys and girls. As per census of 2001-02 (Education Department, J&K) total population of J&K adolescents from 10-14 was 7.9 lakh (4.1 lakh boys and 3.8 lakh girls) and between 15 and 19 years the total population of adolescents was 5.4 lakh (2.5 lakh boys and 2.8 lakh girls). Since the size of adolescents was too large to be covered, it was decided to employ sampling method.

A. Sample Size

A total of 1500 government school students comprising nearly 1% of State's total on roll adolescent population were chosen by Simple Random Sampling. The specific population selected for sampling in the survey was students attending schools from middle to higher secondary.

Sampling Procedure:

The study sample was selected using the following design.

B. Design: - Multi stage sampling procedure.

Thirty sites were selected as follows:

Stage I: Administratively, Kashmir valley at the time of data collection was divided into six districts. Two Tehsils from each district were selected to obtain total of 1500 adolescents aged 10-19 years. All educational Zones falling in each district were enlisted. While using the random tables to obtain total of thirty educational Zones, five Educational Zones were selected from each district

Stage II: The middle, high and higher secondary schools falling in each Tehsil was surveyed for collection of the sample. All the schools i.e. the schools in which the required age group was present were enlisted in each Educational Zone. One school (clustered) was selected in each zone by using.

C. .Random Tables

Stage III: From each District a selected sample of 125 students each both from boys as well as from girls were taken by systemic random sampling as final sampling unit.

D. Data collection

The information was collected from primary as well as secondary sources. In primary sources questionnaire cum interview technique was used. In secondary source journals, books and related literature were studied. In designing questionnaire, simple language was used but still in some schools questions had to be explained in local Kashmiri language to obtain appropriate information from the respondents. Pre-testing was done on 2 per cent of the sample and questionnaire was modified accordingly. The questionnaire consisted of the following sections:

E. General Information

In this section name, address, age, class, parental education, parental occupation, income of the family was asked. Parental literacy is perhaps the most important factor that determines the prevailing state of ill health and under nutrition/malnutrition. It has been observed that educated mothers with inadequate health care and limited economic facilities could largely succeed in escaping ill health and malnutrition because they could utilize available meager resources optimally.

RESULTS

Table 1 Nutritional Deficiencies among Adolescent Boys

Age	No deficiency		Bitot's Spot		Angular Stomatitis		Anemia		Scurvy		x^2
	N	%	N	%	N	%	N	%	N	%	
11-14	197	26	18	2	28	4	66	9	21	3	7.8
15-18	283	38	20	3	42	5	90	12	16	2	1*
Total	480	64	38	5	70	9	156	21	37	5	

*Significant at 0.05

Table No. 1 shows the symptoms of nutritional deficiency of adolescent boys. The clinical assessment of nutritional status shows that 64% of adolescent boys were without any nutritional deficiency. Among the deficiency anemia was very common and it was observed in 21% boys, 5% showed signs of Bitot's spots, 9% angular stomatitis and 5% scurvy. The study findings are comparable to the observations of the Gupta (1973) who reported nutritional deficiencies among adolescents. Present study findings were also better than Milton (1990) who reported 40% of adolescent were anemic. Seltzer(1963) study on adolescents reported that H B level of adolescents was 12 gm/100ml and are slightly better than present studies.

Table 2: Nutritional Deficiencies among Adolescent Girls

Age	No deficiency		Bitot's Spot		Angular Stomatitis		Anemia		Scurvy		x^2
	N	%	N	%	N	%	N	%	N	%	
11-14	180	24	28	4	33	4	125	17	53	7	7.81*
15-18	240	32	11	1	50	7	145	19	67	9	
Total	420	56	39	5	83	11	270	36	120	16	

*Significant at 0.05

Table No. 2 shows the signs of symptoms of nutritional deficiency of adolescent girls. The clinical assessment of nutritional status shows that 56% of adolescent girls were without any nutritional deficiency. Among the deficiencies Anemia was very common and it was observed in 36% of adolescent girls, 5% shows signs of Bitot's spots, 11% angular stomatitis and 16%

scurvy. The findings are better than that of Kurz (1996) who reported 55% of Indian adolescents were anemic. Slemenda (1997) states that most adolescents had nutritional deficiencies like anemia and Vitamin deficiency. Pati (2004) also reported that most of the adolescents were anemic. Carrut and Golberg (1990) also reported nutritional deficiencies among adolescents.

Table 3: Menstrual History of Adolescent Girls

Complaints	Respondents	%age
Irregular	447	60
Excessive	315	42
Scanty	151	20
Painful	241	32
Taking medicine during periods	153	20

Table No. 3 shows menstrual history of the sample. The investigation revealed that 60% adolescent girls had irregular period. The nature of bleeding shows that 42% girls had excessive periods, 20% had scanty periods and 32% had painful periods. The findings of the present study are comparable of Singh (2006) who reported menstrual problems among adolescents. Campbell and Macrath (1997) in his study reported use of medicine during periods by adolescent girls. It can be seen from the above table that measure complaint of adolescent girls was irregular menstrual history i.e. 60%.

Table 4: Distribution of Adolescent by Personal Hygiene

Gender	Very clean		Clean		Unclean		x^2
	N	%	N	%	`N	%	
Boys	419	56	100	13	231	31	35.26*
Girls	481	64	135	18	134	18	
Total	900	60	235	16	365	24	

*Significant at 0.05

Table No. 4 reveal that the personal hygiene of adolescents was satisfactory wherein more than half i.e. 56% of boys and 64% girls were clean. The percentage of unclean adolescent was 31%

boys and 18% girls. The personal hygiene among adolescents was noticed in the form of cleanness of teeth, nails, hair and uniform, mode of brushing, hand clean, bathing and physical appearance and was graded as very clean, clean and unclean. The findings of the present study are comparable with Jalil (1993) who also noticed good personal hygiene of adolescents. Rao (1993) also reported satisfactory hygiene among adolescents.

Table 5:Distribution of Adolescent boys by Height and Weight

Age	Weight in Kgs		Height in Cms		BMI of Boys Mean±SD
	Mean±SD	Recommendation	Mean±SD	Recommendation	
11-14	43±1.41	45	157±1.41	157	17.4±1.38
15-18	44±1.25	65	157±1.41	176	18±1.41

Source: Food & Nutrition Board of India, National Academy of Science
designed for the maintenance of good nutrition

Table No. 5 given above gives distribution of boys by height and weight. It shows mean weight of boys was 43 kgs against recommended 45 kgs and height was 157 cms against recommended 157 cms and BMI was 17.4 in the age group of 11-14 years. The mean weight of adolescent boys was 44 kgs against recommended 65 kgs and height was 157 against 176 and BMI was 18 in the age group of 15-18 years. The findings of the present study are comparable to the findings of Merchant (1975) reported that height and weight of respondents were less as compared to ICMR values. Jain (1973) also reported height and weight of adolescents were less as compared to ICMR values.

Table 6: Distribution of Adolescent girls by Height and Weight

Age	Weight in Kgs		Height in Cms		BMI of Girls Mean±SD
	Mean±SD	Recommendation	Mean±SD	Recommendation	
11-14	42±1.74	46	147±1.78	157	19.4±1.80

15-18	46±1.78	55	153±1.68	160	19.6±1.80

Table No. 6 given distribution of girls adolescents by height and weight. It shows mean weight of adolescent girls was 42 kgs against recommended 46 kgs and mean height was 147 against recommended 157 cms and BMI was 19.4 in the age group of 11-14 years. The mean weight of adolescent girls was 46 kgs against recommended 55 kgs and height was 153 cms against 160 cms and BMI was 19.6 in the age group of 15-18 years. The findings of the present study were better than the survey conducted by National Nutrition Bureau in which it has been revealed that comparatively a high percentage of adolescent girls in the age group of 14-15 years reported with weight 38 kgs and height 145 cms.

Table 7: Age wise distribution of Mean Height and Weight of Adolescent

Age	Examined		Mean Weight		Mean Height	
	N	%	Kgs	±SD	Cms	±SD
11	134	9	40	2.73	151	2.31
12	200	13	42	3.00	153	2.57
13	220	15	43	5.50	154	2.59
14	262	17	43	4.00	155	3.53
15	175	12	45	1.22	156	1.83
16	167	11	45	4.70	158	4.27
17	177	12	48	2.5+	158	4.67
18	165	11	46	1.41	159	1.73

Table No.7 given the age wise distribution of mean height and weight of the adolescents. The mean weight and height of 11-year-old adolescent was 40 kgs. and 151 cms. respectively. The mean weight and height of 12-year-old adolescent was 42 kgs. and 153 cms. respectively. The mean weight and height of 13-year-old adolescent was 43 kgs. and 154 cms. The adolescents in the group of 17 years showed maximum weight i.e. 48 kgs and 158 cms. height. These findings are comparable to the findings of Koshi (1970) who reported that mean height and weight of the adolescents especially girls was less as compared to the ICMR standard values. Bhandari (1975) in the study of health and nutrition of 839 schoolboys of Udaipur aged 5 to 17 years found that the mean height and weight was less as compared to the ICMR values and international standards.

Table 8: Age wise distribution of Adolescents Boys by BMI Categories

Age	BMI <18.5 under weight		MBI 18.5<25 Normal weight		BMI 25<30 Over weight		BMI >30 obese	
	N	%	N	%	N	%	N	%
11	38	42	29	32	23	26	*	*
12	40	41	33	34	19	19	6	6
13	47	46	41	40	12	12	3	2
14	41	38	36	34	22	21	8	7
15	36	38	34	36	18	19	7	7
16	31	40	23	30	18	23	5	7
17	20	22	43	46	25	27	5	5
18	29	33	27	31	20	23	11	13

Table No.8 gives the age-wise distribution of adolescent by BMI. It was found that 42% of adolescent boys were under weight with BMI l < 18.5. 32% boys were normal with BMI more than 18.5 < 25. 26% of them were overweight with BMI more than 25<30 in the age of 11 years .41% of boys were under weight, 34% Normal weight, 19% over weight, 6% obese in the age of 12 years .46% of adolescents were under weight 40% of normal 12 % over weight and 2% obese in the age of 13 years. 38% of boys were under weight, 34% normal weight, 21% overweight, 7% obese in the age of 14 years.38% of boys were under weight, 36% normal weight 19% overweight and 7% obese in the age group of 15 years. 40% boys were under weight, 30% normal weight`, 23% overweight and 7% obese in the age of 16 years. 22% of adolescent were under weight, 46% normal weight, 27% overweight and 5% obese in the age of 17 years. 33% of boys were under weight and 13 were obese in the age of 18 years. The observations in the present study were better than reported findings of Hampton (1967 and 1966), who reported 14% of boys and girls were graded as obese and in another study he reported 20% of boys and 25% girls were under weight were Desmukh (1992) also reported that 53.8% of adolescent were thin, 4% normal and 2.2% were overweight.

Table 9: Age wise Distribution of Adolescent Girls by BMI Categories

Age	BMI <18.5 under weight		MBI 18.5<25 Normal weight		BMI 25<30 Over weight		BMI >30 obese	
	N	%	N	%	N	%	N	%
11	21	48	16	36	5	11	2	5
12	37	36	43	42	19	19	3	3
13	36	31	49	42	22	19	10	8
14	49	32	56	36	39	25	11	7
15	31	39	27	34	18	22	4	5
16	34	27	38	42	20	22	8	9
17	35	42	29	35	16	19	4	4
18	29	37	33	42	13	17	3	4

Table No.9 gives age wise distribution of adolescent girls by BMI. It was found that 48% girls were under-weight. 36% were normal weight, 11% were overweight. 5% obese in the age of 11 years, 36% girls were underweight, 42% normal weight, 19% overweight, 3% obese in the age of 12 years, 31% girls were underweight, 42% normal weight, 19% overweight and 8% obese in the age of 13 years. 32% girls were under weight, 36% normal weight, 25% overweight and 7% obese in the age of 14 years. 39% girls were underweight, 34% normal weight, 22% overweight and 5% obsess in the age of 15 years, 27% of girls were underweight, 42% were normal weight, 22% were overweight, 9% obese in the age of 16 years. 42% girls were underweight, 35% normal weight, 19% overweight and 4% obese in the age of 17 years. 37% girls were underweight, 42% were normal weight, 17% overweight and 4% obese in the age of 18 years. The findings of the present study can be compared with the findings of Sinhababu who worked on 176 students of Nursing Training school of Bankura July 2006 and reported that the proportion of overweight was only 5.1% among students. The proportion of underweight was 33.5% and reported BMI of adolescents 20.24, 19.97 and 19.4 in the three groups of adolescents.

Table 10: Distribution of Adolescents by Body Appearance

Gender	Average		Large		Small		x^2
	N	%	N	%	N	%	
Boys	378	50	95	13	277	37	11.56*
Girls	419	56	115	15	216	29	

Total	797	53	210	14	493	33	

*Significant ay 0.05

Table No.10 shows the distribution of adolescent by body appearance. The maximum number of adolescent i.e. 50% boys and 56% girls were in average body frame. 13% boys and 15% girls were in large frame. 37% boys and 29% girls were in small frame. Different studies on Nutritional status of adolescent girls and boys conducted in Mumbai, Kolkata and Chennai by WHO 1998 revealed that comparatively a higher proportion of adolescent girls were found victims of malnutrition because of poor nutrition that reflected their poor growth, abnormal body size and shorter height.

Table 11: Nutritional Intake of Adolescent Boys by Age as per 24 Hours Diet Recall

Nutrients	Age	
	11-14	15-18
	Mean±SD	Mean±SD
Protein (Grams)	54±1.41	51±1.42
Calories (K.Cal.)	2350±1.42	2028±1.43
Fat (Grams)	44±1.71	47±1.41
Ret . (Iu)	420±1.48	413±1.39
Carotene (Iu)	2430±1.42	2090±1.42
Calcium (Grams)	0.48±1.42	0.52±1.46
Iron (mg)	11.9±1.87	12.6±1.81
Vitamin B1 (mg)	0.9±1.93	1.2±1.85
Vitamin B2 (mg)	1.3±1.89	1±1.73
Vitamin B3 (mg)	24.7±1.80	23.8±1.78
Vitamin C (mg)	49.8±2.0	43.03±1.77

Table 12: Nutritional Intake of Adolescent Boys by Age as per 24 Hours Diet Recall

Nutrients	Age	
	11-14	15-18
	Mean±SD	Mean±SD
Protein (Grams)	48.2±1.79	50.03±1.78
Calories (K.Cal.)	2166±1.43	2094±1.42

Fat (Grams)	40.8±1.78	41.2±1.68
Ret . (Iu)	308±1.77	329±1.77
Carotene (Iu)	2259±1.77	2102±1.78
Calcium (Grams)	0.46±1.76	0.49±2.26
Iron (mg)	10.7±1.73	11.8±1.79
Vitamin B1 (mg)	1.1±2.03	1.0±1.77
Vitamin B2 (mg)	1.1±2.35	1.05±1.41
Vitamin B3 (mg)	22.3±1.38	21.9±1.40
Vitamin C (mg)	40.8±1.78	39.1±1.79

Table 13: Consumption Pattern of Fleshy Foods by Adolescents (N=750 each)

Gender	Food Items	Daily		Twice in a week		Weekly		Rarely		x^2
		N	%	N	%	N	%	N	%	
Boys	Meat	313	42	219	29	210	28	8	1	
	Organ Meat	212	28	150	20	345	46	44	6	
	Poultry	83	11	179	24	315	42	173	23	
	Fish	79	11	167	22	287	38	219	29	
	Eggs	404	54	274	37	66	9	6	0.8	43.67*
Girls	Meat	357	48	229	30	126	17	38	5	
	Organ Meat	301	40	157	21	267	36	25	3	
	Poultry	69	9	285	38	328	44	68	9	
	Fish	85	11	193	26	267	36	205	27	
	Eggs	375	50	306	41	53	7	16	2	

*Significant 0.05

Table No.13 shows the consumption pattern of fleshy food by adolescents. The number of girls who consume meat daily was 48% and boys were 42%. 54% of boys and 50% of girls consumed eggs daily. 28% of boys and 40% of girls consumed organ meat daily. 11% of boys and 11% of girls consumed fish daily. 11% of boys and 9% girls consumed poultry daily. The findings are comparable to the findings of Kurz (1996) who reported because of less consumption of fleshy food which are rich sources of Iron, many adolescents develop Iron deficiency anemia. Pati

(2004) a study on adolescent girls also reported poor intake of fleshy foods by adolescent because of low socio status.

Table 14: Consumption Pattern of Pulses and Dal Products by Adolescents(N=750 each)

Gender	Daily		Twice in a week		Weekly		Rarely		x^2
	N	%	N	%	N	%	N	%	=31.35
Boys	224	30	375	50	119	16	33	4	*
Girls	301	40	239	44	113	15	7	0.9	

*Significant 0.05

Table No.14 shows the consumption pattern of pulses and *dals* by adolescents. The consumption of *dals* by adolescents is seasonal in Kashmir. They mainly eat them during winter only. 30% boys and 40% girls consumed *dals* daily. 50% boys and 44% girls consumed it twice in a week. 16% of adolescent boys and 15% of adolescent girls consumed pulses and dal products weekly. 4% of adolescent boys and 0.9% adolescent girls consumed pulses and dals rarely.

Table 15: Consumption Pattern of Milk and Milk Products by Adolescent (N=750 each)

Gender	Food Items	Daily		Twice in a week		Weekly		Rarely	
		N	%	N	%	N	%	N	%
Boys	Milk	367	49	235	31	90	12	58	8
	Curds	400	53	212	28	80	11	58	8
	Butter	329	44	232	31	139	18	50	7
Girls	Milk	340	45	245	33	86	11	79	11
	Curds	355	47	219	29	69	9	107	14
	Butter	298	40	219	29	146	19	87	12

Table No.15 gives the consumption pattern of milk and milk products by adolescents. 49% and boys and 45% of girls consumed milk daily. 53% of boys and 47% of girls consumed curds daily. 44% of boys and 40% of girls consumed butter daily. The present study findings are comparable to the findings of Davis (1969) an extensive research on adolescent food habits and nutritional status revealed that consumption of milk and milk products were very less in adolescent diet because of poverty.

Table 16: Consumption Pattern of Fruits by Adolescent (N=750 each)

Gender	Daily		Twice in a week		Weekly		Rarely		x^2
	N	%	N	%	N	%	N	%	=11.39
Boys	445	59	165	22	130	17	10	1.4	*
Girls	423	56	167	22	129	17	31	4	

*Significant 0.05

Table No.16 shows the consumption pattern of fruits by adolescents. 59% boys and 56% of girls consumed fruits daily. 22% of boys and 22% girls consumed fruits twice in a week and 17% of boys and girls consumed fruits weekly. 1.4% of boys and 4% of girls consumed fruits rarely. The consumption of fruits is seasonal in Kashmir. The consumption patterns of fruits showed almost more than half percentage of adolescents enjoyed eating fruits daily. This is because most of the Kashmiri fruits were grown locally and adolescents eat them with great joy.

Table 17: Consumption Pattern of Dry Fruits by Adolescent (N=750 each)

Gender	Daily		Twice in a week		Weekly		Rarely		x^2
	N	%	N	%	N	%	N	%	=9.25*
Boys	250	33	150	20	200	27	150	20	
Girls	225	30	120	16	220	29	185	25	

*Significant 0.05

Table No.17 shows the consumption pattern of dry fruits by adolescents. 33% boys and 30% of girls consumed dry fruits daily. 20% of boys and 16% girls consumed it twice in a week. 27% of boys and 29% of girls consume other it weekly. 20% of boys and 25% of girls consume it rarely. The consumption pattern revealed that only one third of boys and 30% girls eat dry fruits daily. This is because only few types of dry fruits are locally grown in Kashmir and others to be imported from other states. As most of the adolescents belong to poor and low economic status family, it is not under their budget and cannot therefore afford.

Table 18: Distribution of Adolescent by Food Preferences (N=750 each)

Gender	Hot Spicy Food	Less Spicy Food	Grilled	Roasted	Fried	x^2

	N	%	N	%	N	%	N	%	N	%	=27.13
Boys	418	56	75	10	105	14	95	13	87	12	*
Girls	475	63	61	8	128	17	447	60	513	68	

*Significant 0.05

Table No.18 shows distribution of adolescents by food preferences. The maximum number of adolescents i.e. 56% and 63% girls liked to have hot spicy foods. 10% of boys and 8% girls liked less spicy foods, 14% boys and 17% girls liked grilled foods. 13% boys and 60% of girls liked roasted foods, 12% boys and 68% girls liked fried foods.

Table 19: Water Habits of Adolescents (N=750 each)

Gender	Boiled/ filtered water		Boiled/unfiltered water		x^2
	N	%	N	%	=55.85
Boys	400	53	350	47	*
Girls	540	72	210	28	

*Significant 0.05

Table No.19 shows the water habits of adolescents. While analyzing the table it was observed that still 47% of boys and 28% of girls use un-boiled and un-filtered drinking water. The report of the findings was in contrast to the report of Economic Survey 2006-07 that Safe drinking water was not available to 34.8% of households.

Table 20: Mass Media Temptation of Adolescents (N=750 each)

Gender	Temptation		Non-Temptation		x^2
	N	%	N	%	
Boys	380	51	370	49	=5.42*
Girls	425	57	325	43	

*Significant 0.05

Table No.20 shows the number of boys who showed temptation towards mass media is 51% and girls 57%. Adolescents being the keen observers of the TV and other mass media sources they were tempted by foods adds and prefer them in their daily routine. This affects their diet and adolescents were victims of most dreaded diseases.

DISCUSSION

Nutritional Deficiencies among Adolescent: The clinical assessment of nutritional status shows that 48 i.e. 64 per cent of boys and 5.6 per cent of girls were without any deficiency disease. Among the deficiencies anemia was common and was observed in both sexes. 21 per cent of boys and 36 per cent of girls showed signs of anemia. The mean HB level of anemic girls was 9 gms out of 36 per cent of anemic girls 7.6 per cent had severe anemia with HB level of 7 gms. In boys, the mean HB level of anemic boys was 10.5 gms. 5 per cent of boys and 5 per cent girls shows signs of Bitot's spots. 9 per cent of boys and 10 per cent of girls shows signs of scurvy. When nutritional deficiencies of adolescents were compared with their parental income, more nutritional deficiencies were observed in lower income families. The nutritional deficiencies were more seen in urban adolescent than rural adolescent because of easy availability of food items and also they eat meat of other domestic animals which are cheap as compared to sheep in urban area. Because of lower calorie requirement the iron intake is likely to be no more than 10-12 mg/day even though the other aspects of the diet may be fully adequate with absorption at 5-10 per cent, there intakes are not adequate to cover fully the losses entailed by menstruation and to build up reserves. Adolescent girls in addition must have iron to meet their growth requirement yet during these years their diets are often of poor quality and quantity. This causes anemia to the vast majority of adolescent which intern becomes the breeding ground for generation of other diseases. So to prevent the deficiencies adolescents should have enough quantity of micronutrients especially iron and vitamins. Vitamin C which promotes the absorption of iron, is also deficient in the diets of adolescents in both rural and urban adolescents because of low socio economic status. A comprehensive study conducted in twin cities of Hyderabad and Secandrabad (Nagraj, 1975) in school children with regard to physical, nutritional and dental health reported that among 1200 children in the age of 5-15 years 40 per cent had dental caries, 8 per cent eye disorders, 5 per cent anemia, 4.5 per cent signs of vitamin B complex deficiency and 4 per cent signs of Vitamin A deficiency. The findings of the present study were somewhat better than WHO report (1998) where it is reported that 40-60 per cent of anemia is seen among adolescent. The study findings are comparable to the observations of Gupta (1973) who reported nutritional deficiencies among rural and urban adolescent. Present study findings were also better than Milton (1990) who reported 40 per cent of adolescent were anemic. Seltzer (1963) study on adolescent reported that HB level of adolescent was 12 gm/100 ml and slightly better than the present studies. Kurz (1996) reported 55 per cent of Indian adolescent are anemic. Slemenda (1997) states that most adolescent had nutritional deficiencies like anemia and Vitamin deficiency. Pati (2004) also reported that most of adolescent girls were anemic. The findings of a survey conducted by Federal Centers for disease control and prevention USA (1998) revealed that nearly 8 million young woman are deficient in iron. Carruth and Golberg (1990) also reported nutritional deficiencies among adolescents. These findings highlight the need for organizing a very special type of nutrition education program to correct the existing imbalances. The poverty coupled with non-affordability of nutritious foods give rise to health related problems among significant sections of the adolescent. The availability of good foodstuffs is main source of nutrition in rural areas, which are not available for urban people without proper payment. Moreover, the vegetables which are in rural areas consumed a fresh growing gives good nutritive returns in comparison to urban localities were stored and preserved items are used. Furthermore the availability of fresh vegetables and fruits are comparatively more expensive which rarely permits the purse of low class people in urban areas. So to prevent the deficiencies and correct the imbalances adolescent should be provided free lunch in schools as part of supplementary food.

Nutritional Status of Adolescent: More and more evidence suggests that lifestyle including nutritional habits track from adolescent into adulthood, thereby lead to increased incidence of chronic disease e.g. cardiovascular diseases, diabetes and cancer. Exposure in childhood and adolescence to adverse life style and faulty food habits such as poor food intake, special meal preferences and patterns and above all sedentary life style may exacerbate this, thus worsening the prognosis. Hence promotion of healthy nutrition habits and physically active life style during adolescent period is a critical public health strategy.

CONCLUSION

The age wise distribution of adolescents revealed 44 per cent were in the age group of 13-15 years, 34 per cent in the age group of 16-18 years and 22 per cent in 10-12 years. 55 percent of boys and 54 per cent of girls were from illiterate families.

> ➤ The maximum number i.e. 50 per cent of adolescent boys were from service class families and similarly maximum girls i.e. 40 per cent were from skilled labour family. 57

per cent of boys and 60 per cent of girls were from joint families. The survey revealed that 30 per cent of boys and 5 per cent girls were in BPL category. 20 per cent of girls and 15 per cent boys were in low-income group i.e. Rs.450-860 income group. 51 per cent of boys and 34 per cent of girls were from Rs.860- 1460 income group.

➢ The personal hygiene of adolescents was satisfactory with only 31 per cent boys and 18 per cent of girls were unclean. The morbidity pattern of adolescents' shows 29 per cent of boys and 34 per cent of girls complained of burning sensation of eyes. 24 per cent of boys and 24 per cent of girls have difficulty in reading and 51 per cent of girls and 43 per cent of boys showed dental cavities. 44 per cent of boys and 34 per cent girls were found with inflamed tonsils. 10 per cent of boys and 15 per cent of girls complained of breathlessness.

➢ The consumption pattern of adolescents' shows that only 59 per cent of boys and 56 per cent of girls eat fruits daily. 42 per cent of boys and 48 per cent of girls eat protein rich foods like meat-chicken-fish-egg daily. 48 per cent of boys and 45 per cent of girls consume dairy products daily, 30 per cent of boys and 40 per cent of girls consume pulses and dals daily but the consumption is seasonal only in winter.

➢ The mean intake of adolescents showed that they were consuming less as compared to balanced diet standards. The mean consumption of cereal food like rice ranges from 192.5– 200 gmsfor boys & 150 – 200 gms for girls. The average intake of green leafy vegetables for boys varies from 62-83 gms and for girls 58-75 gms. The average intake of protein food like fleshy food for boys varies between 26.1–29.3 gms and for girls 25– 27.8 gms. The consumption of dairy products like milk varies from 95–122 ml for boys & 72.5–126 ml for girls. The intake of dals varies from 42.16 –47.6 gms and for girls 44.5– 49.6 gms.

REFERENCES

1) Adams J.F. (1973). Understanding Adolescence. Current Developments in Adolescent Psychology. 2nd edition.

2) Antia F.P. (1973) Clinical Dietetics & Nutrition. London: Oxford University Press.

3) Apley J. (1975). The Child with Abnormal Pains. Oxford Black Wills Scientific Publications.

4) CarinoVereeken and Leas Macs (1992) Adolescent Eating Habits, Dental Care and Dieting. Health Behavior of School Aged Children. British Journal of Nutrition, 4.

5) Chant, S. (1997). Woman Headed Households: Diversity and Dynamics in Developing World Basingstoke- Macmillon Press.

6) Chaplin, J.P. (1982) Dictionary of Psychology. New York: Dell.

7) Chariton A and Blair V (1989). Absence from School and Drug Use Related to Children's and Parental Smoking Habits. British .289:90-92

8) Chatterji Satipati (2002) A Longitudinal Study on Sports and Exercise, A Medical Journal, Department of Physiology, university of Science & Technology, Kolkata - India.121(1)35 .

9) Diet & Health in School Aged Children (1995) London Health Education Authority.

10) Dorland (1974). Illustrated Medical Dictionary 25th Ed. London: W.B. Saunders.

11) Dreyer (1982) Sexuality during adolescence. Hand Book of Developmental Psychology, B. B .Wolman .569-601.

12) Erikson, E. H. (1964). Childhood and Society. Revised Edition New York: Norton.

13) FAO/WHO/UNO Energy & Protein Requirement Report of Joint (1985). Expert Consultation WHO Technical Report Series Geneva WHO. 724.

14) Field, T (1995). Adolescents Intimacy with Parents and Friend. Adolescence – 30. (117)113-140.

15) Fleek, H (1981). Introduction to Nutrition 4th. Edition. New York: MacMillan Publishing Company.

16) Fleming, C M (1948). Adolescent & its Social Psychology. London: Routledge and Kegan Paul Carter Lane London.

17) Fletcher, I. (1963). Anxiety & Achievement of Intellectually Gifted & creative Children. Journal of Psychology. 56:167- 170.

18) Foster (1997). Community Health Nursing Theory & Practice 2nd. Edition W.B. Saunders Company.

19) French, Joan (2001). Gender Equality and Rights of Woman and Girls, Development Vol. 44 Crom Well Press, Wiltshire.

20) Furstenberg, Frank F. (1987). Race Differences in Teenage Sexuality, Pregnancy & Adolescent Child Bearing. New York: Milbank Quarterly 65.

21) Garbarino J. &Sebes J. (1984). Families of Risk for Destructive Parent Child Relations in Adolescent. Child Development. Pub .Med .Gov .U S National Library of Medicine, 55:174-183.

22) Garg B.S. (Sep.2002) Indian Journal of Population Education Indian. Adult education Association National Documentation Center on literacy & Population Education In .jour. of pop.ed .-18

23) Goodhart, Robert S & Shills, Maurice E. (1980). Modern Nutrition in Health & Disease. 6th Edition. Great Britain: Henry Kampton.

24) Gopalan, C. (1968). Nutritive value of Indian foods. National Institute: of Nutrition Indian Council of Medical Research Hyderabad, India.

25) Gopal, A. K. (1975). Certain Differentiating Personality Variables of Creative & Non-creative Science & Engineering Students ph .D Edil .Kur .u -2.

26) Gour, Sood and Gupta A.K. (1989). Goiter in School Girls of Mewat Area Haryana. Indian Pediatrics, 26:223-227.

27) Govt. of J & K Digest of Statistics 2000-01 and 2001-02. Directorate of Economics & Statistics Planning & Development Department.

28) Groves D. (1988). Is childhood obesity related to T V. Addiction. Physician and sports medicine 16(11)116-118, 120-122.

29) The Common Wealth Fund, Health Concerns Across a Woman's Lifespan (1998). Survey of Woman's Health.

30) Wiehl, D G (1941). Selected Cases of Anemia Among Adolescents. American Journal of Public Health. 31(10)1073- 1078.

31) Wilkinson, R (1986). Income and mortality in Adolescents. London: Health Research & Longitudinal Data Tavistock.

32) William, C L (1995). Importance of Dietary Fiber in Childhood. Journal of American Dietetic Association. 95(10)1140- 1149.

33) Woman Health & Development (1985). WHO off set Publication No. 90 A Report by Director General WHO Geneva.4-11.

34) Wood Ward David R. (1985). What Sort of Teenager has Low Intakes of Energy Nutrients? British Journal of Nutrition.53:241-249

35) World Health Statistics (1987) Annual Report WHO

36) Zelnik M .and Kantrer J. (1977) Sexual and contraceptive experience of young unmarried women in the United States, Family Plann, Prospect 9 (2).